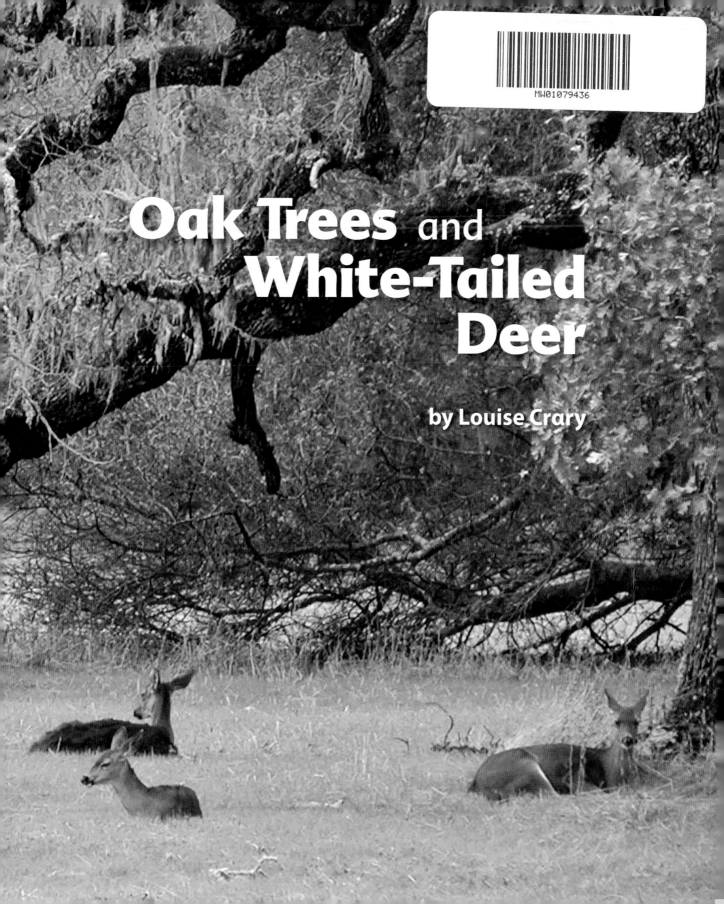

# Oak Trees and White-Tailed Deer

by Louise Crary

# Contents

# Science Vocabulary

## root

A **root** is the part of a plant that takes in nutrients and water from the soil.

An oak tree's **roots** grow in the soil.

## stem

A **stem** is the part of a plant that carries water and nutrients to the leaves and food back to the roots.

An oak tree's trunk is its **stem.**

## leaf

A **leaf** is the part of a plant that makes food for the plant.

Each oak tree **leaf** takes in sunlight and air to make food.

## flower

A **flower** is the part of a plant that makes seeds.

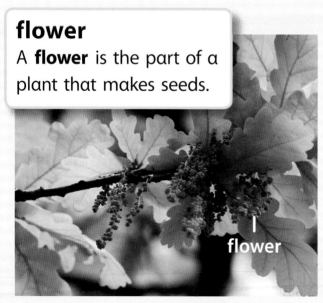

flower

Oak trees have **flowers.**

## seed

A **seed** is a part of a plant from which another plant can grow.

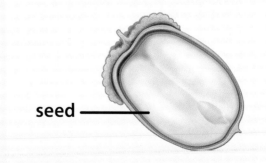

seed

**Seeds** grow inside acorns.

5

## life cycle

A **life cycle** is the way a living thing grows, changes, makes more living things like itself, and dies.

**Fawn**

**Young deer**

**Adult deer**

A white-tailed deer grows and changes during its **life cycle.**

6

## offspring

The young of a plant or animal is its **offspring**.

A fawn is a deer's **offspring**.

## trait

A **trait** is a feature or behavior passed on from parents.

A white tail is a **trait** of this kind of deer.

**My Science Vocabulary**

flower

leaf

life cycle

offspring

root

seed

stem

trait

# Parts of an Oak Tree

Many trees grow in this forest. Some are oak trees. Other kinds of plants grow in the forest, too.

Animals live in the forest. They build homes and find food in the forest trees.

This forest is in a national park in Virginia.

All plants have parts. Each part does a different job. An oak tree's parts help it stay alive. Some oak tree trunks are much larger than many other plants' **stems.**

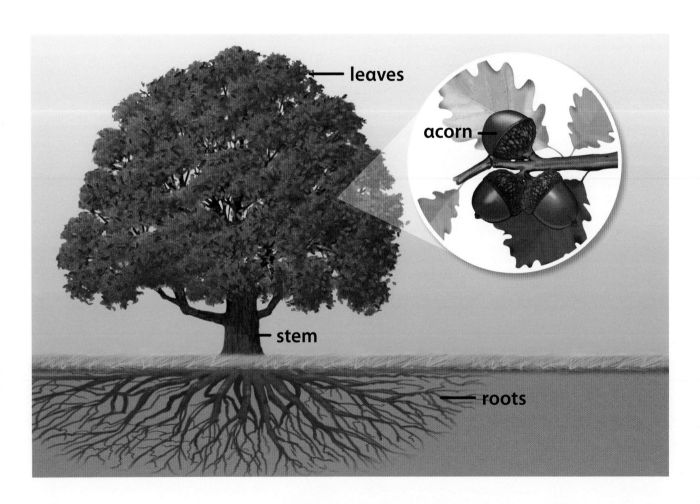

leaves

acorn

stem

roots

**stem**

A **stem** is the part of a plant that carries water and nutrients to the leaves and food back to the roots.

Oak tree **roots** grow deep and wide. They take in water and nutrients to help the tree grow.

Deep and wide roots hold the tree in place in the ground.

**root**

A **root** is the part of a plant that takes in nutrients and water from the soil.

An oak tree's stem is its trunk. Rough, hard bark covers the trunk. The bark protects the water and nutrients inside the trunk.

As an oak tree grows, its trunk gets thicker.

Each oak tree **leaf** takes in sunlight to help the plant make food. In summer, there is more sunlight to keep leaves green.

With less sunlight, oak tree leaves turn yellow and fall off the tree.

In summer, leaves make food that helps the tree stay alive all year.

**leaf**

A **leaf** is the part of a plant that makes food for the plant.

**Flowers** grow on oak trees. Over time, some of the flowers make fruit. The fruit of an oak tree is an acorn.

— flower

**flower**

A **flower** is the part of a plant that makes seeds.

Inside most acorns is an oak tree **seed.**
In the fall, acorns drop from oak trees.
Some oak tree seeds might begin to
grow in the soil.

acorn

The white part inside an acorn is the seed.

**seed**

A **seed** is a part of a plant from
which another plant can grow.

# An Oak Tree Life Cycle

The seed inside an acorn is the beginning of an oak tree's **life cycle**. A life cycle is the way a living thing grows and changes.

**Adult tree**

**life cycle**

A **life cycle** is the way a living thing grows, changes, makes more living things like itself, and dies.

**Seed**

**Seedling**

**Sapling**

# Traits of White-Tailed Deer

Different kinds of animals live in forests.

White-tailed deer are large forest animals.

White-tailed deer look for grasses and twigs to eat during winter.

White-tailed deer have fur that keeps them warm. White-tailed deer also have strong legs for running.

A white-tailed deer's fur becomes thicker in the fall.

White-tailed deer share some of the same **traits,** or features. Deer get these traits from their parents. A white-tailed deer uses its tail to signal other deer.

The white fur on the deer's tail stands out against the brown grass.

**trait**

A **trait** is a feature or behavior passed on from parents.

Male and female deer have some different traits. Male deer, or bucks, have antlers. Female deer, or does, do not.

antlers —

This buck will shed its antlers at the end of the fall. New antlers grow each summer.

In the spring or summer, does give birth to fawns, or newborn deer. A fawn is the **offspring** of deer. It has many of the same traits as its parents do.

A fawn is smaller than a doe.

**offspring**

The young of a plant or animal is its **offspring.**

Fawns have different fur. Their fur has patches of white. As they grow, the patches disappear.

The patches of white fur on a fawn help it blend into the forest.

# A White-Tailed Deer Life Cycle

As a fawn grows, it begins to look more like its parents. Its fur changes. Its ears get bigger.

**Adult deer**

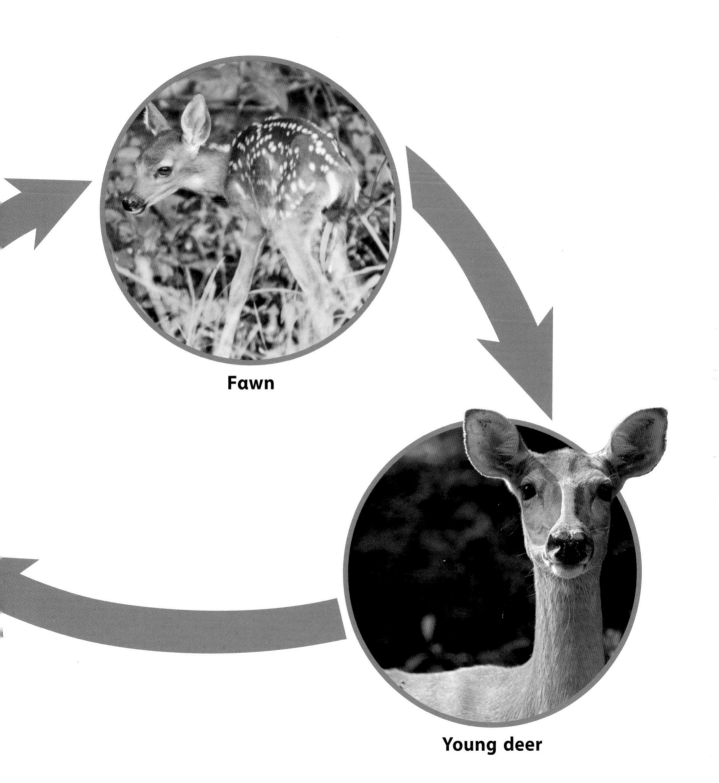

**Fawn**

**Young deer**

White-tailed deer grow and change near forests. Some forests have oak trees.

Like other plants and animals, oak trees and white-tailed deer have life cycles. A white-tailed deer is born live. An oak tree begins its life as a seed.

This white-tailed deer finds the food it needs in the forest.

# Conclusion

There are many plants and animals in a forest. The oak tree and white-tailed deer are two examples. These living things can grow and change in a forest.

## Think About the Big Ideas

1. How are oak trees alike and different from other plants?
2. How are white-tailed deer alike and different from other animals?
3. How do oak trees and white-tailed deer change during their life cycles?

# Share and Compare

 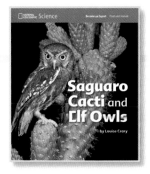 

## Turn and Talk

Compare the plants and animals in your books.
How are they alike? How are they different?

## Read

Find your favorite photo and
read the caption to a classmate.

## Write

Describe the traits of the animal in your book.
Share what you wrote with a classmate.

## Draw

Draw the plant from your book and label its parts.
Share your drawing with a classmate.

## Meet Greg Marshall

Scientists use tools and technology. These tools can help them study animals.

Greg Marshall needed a special camera to study animals. So he invented Crittercam. Crittercam records animals' behaviors in the wild.

# Index

**Acknowledgments**

Grateful acknowledgment is given to the authors, artists, photographers, museums, publishers, and agents for permission to reprint copyrighted material. Every effort has been made to secure the appropriate permission. If any omissions have been made or if corrections are required, please contact the Publisher.

**Photographic Credits:**

Cover (bg) Glenda M. Powers/Shutterstock; Cvr Flap (t), 16 (r) Danita Delimont/Alamy Images; Cvr Flap (c), 18 Stephen Krasemann/NHPA Limited; Cvr Flap (b), 7 (t), 22 TBD PureStock/SuperStock; Title (bg) Mike Liu/iStockphoto; 2-3, 28 Raymond Gehman/National Geographic Image Collection; 4 (t), 11 Adrian T Jones/Shutterstock; 4 (b), 12 Konstantin Baskakov/Shutterstock; 5 (t), 17 (c) Arco Images GmbH/Alamy Images; 7 (bl), 14 Shironina Lidiya Alexandrovna/Shutterstock; 6 (t), 25 (t) Corel; 6 (bl), 24 (r) Dennis Donohue/Shutterstock; 6 (br), 25 (b) Condor 36/Shutterstock; 7 (b), 20 Alan G. Nelson/Animals Animals; 8-9 Owaki - Kulla/Corbis; 13 (l) Flashon Studio/Shutterstock, (r) vnlit/Shutterstock; 15 (inset) Guy Erwood/Shutterstock; 16 (l) Arco Images GmbH/Alamy Images; 17 (b) PhotoArt UK/Alamy Images; 19 Corel; 21 Alan D. Carey/PhotoDisc/Getty Images; 23 Corel; 24 Robert Ranson/Shutterstock; 26 Tom Tracy Photography/Alamy Images; 27 Mark Raycroft/Minden ; Pictures; 30-31 (insets) National Geographic Remote Imaging; 31 Mark Thiessen/National Geographic Image Collection; Inside Back Cover (bg) Mares Lucian/Shutterstock.

**Illustrations:**

10 Miro Design, 15 Paul Mirocha

Neither the Publisher nor the authors shall be liable for any damage that may be caused or sustained or result from conducting any of the activities in this publication without specifically following instructions, undertaking the activities without proper supervision, or failing to comply with the cautions contained herein.

**Published by National Geographic School Publishing & Hampton-Brown**

Sheron Long, Chairman
Samuel Gesumaria, Vice-Chairman
Alison Wagner, President and CEO
Susan Schaffrath, Executive Vice President, Product Development

**Editorial:** Fawn Bailey, Joseph Baron, Carl Benoit, Jennifer Cocson, Francis Downey, Richard Easby, Mary Clare Goller, Chris Jaeggi, Carol Kotlarczyk, Kathleen Lally, Henry Layne, Allison Lim, Taunya Nesin, Paul Osborn, Chris Siegel, Sara Turner, Lara Winegar, Barbara Wood

**Art, Design, and Production:** Andrea Cockrum, Kim Cockrum, Adriana Cordero, Darius Detwiler, Alicia DiPiero, David Dumo, Jean Elam, Jeri Gibson, Shanin Glenn, Raymond Godfrey, Raymond Hoffmeyer, Rick Holcomb, Cynthia Lee, Anna Matras, Gordon McAlpin, Melina Meltzer, Rick Morrison, Cindy Olson, Christiana Overman, Andrea Pastrano-Tamez, Leonard Pierce, Cathy Revers, Stephanie Rice, Christopher Roy, Janet Sandbach, Susan Scheuer, Margaret Sidlosky, Jonni Stains, Shane Tackett, Andrea Thompson, Andrea Troxel, Ana Vela, Teri Wilson, Brown Publishing Network, Chaos Factory, Inc., Feldman and Associates, Inc.

**The National Geographic Society**

John M. Fahey, Jr., President & Chief Executive Officer
Gilbert M. Grosvenor, Chairman of the Board

**Manufacturing and Quality Management, The National Geographic Society**

Christoper A. Liedel, Chief Financial Officer
George Bounelis, Vice President

National Geographic School Publishing
Hampton-Brown
P.O. Box 223220
Carmel, California 93922
www.NGSP.com

Printed in the USA.

ISBN: 978-0-7362-5560-8

10 11 12 13 14 15 16 17

10 9 8 7 6 5 4 3 2 1